TRACKSIDE MARYLAND

FROM RAILYARD
TO MAIN LINE

Just above the Patapsco River valley one humid summer evening, Jim Gallagher caught the mood of the Old Main Line as it curves from Relay to Ellicott City, Maryland. He called this photo "Rails Bend West."

TRACKSIDE MARYLAND

FROM RAILYARD TO MAIN LINE

Photographs by James P. Gallagher
Text by Jacques Kelly

The Johns Hopkins University Press
Baltimore and London

Originally published in 1992 by Greenberg Publishing Company, Inc., Sykesville, Maryland
Johns Hopkins edition, 2003
2 4 6 8 9 7 5 3 1

The Johns Hopkins University Press
2715 North Charles Street
Baltimore, Maryland 21218-4363
www.press.jhu.edu

Library of Congress Cataloging-in-Publication Data

Gallagher, James P.
Trackside Maryland : from railyard to main line /
photographs by James P. Gallagher ;
text by Jacques Kelly.
p.   cm.
Originally published: Sykesville, MD : Greenberg Pub. Co., c1992.
ISBN 0-8018-7323-1 (alk. paper)
1. Railroads—Maryland. I. Kelly, Jacques, 1950–   II. Title.

TF24.M3 G35 2003
385′.09752—dc21   2002028254

A catalog record for this book is available from the British Library.

A southbound Ma & Pa extra crosses the Gross Trestle at Sharon, Maryland, 1955.

CONTENTS

The Photographer

Jim Gallagher, a member of the famous 49th Fighter Group, was among the first occupation forces in Japan after its surrender. He brought back a rabbit fur-lined flying suit found in a Japanese military base. The war outfit served him well. Here he has taken a self-portrait posed with his trusty "beat" Graphic. His Rolliflex, set to make the timed shot, stands some feet away.

INTRODUCTION

It was a Sunday morning in December 1955. I can well recall my grandmother's kitchen, the scent of the scrapple and flannel cakes mixed in with the after-church family conversation. My five-year-old eyes were fixed on *The Sunday Sun Magazine*. To me, gazing at James Gallagher's photo essay of the Maryland & Pennsylvania Railroad was the most sacred event of that day. Come nightfall, I had cut and pasted those sepia-toned pictures in my scrapbook. And, after thirty-seven years, I have never lost that set of the Ma & Pa's No. 27 scenes.

Years later, I admired Jim's classic silhouette of a Ma & Pa steam train on an old Harford County trestle. Without telling me, my mother contacted Jim directly and made a request for a print. She presented me with a fine 11 x 14 inch photograph one Christmas morning. That photograph still hangs in my dining room.

That same Ma & Pa photo later turned up on the editorial pages of *The Evening Sun*. It bore the credit line, "Photo by James P. Gallagher." I called Jim on the phone and we met. As luck would have it, our offices were then but a city block apart. He was a stockbroker at Robert Garrett and Sons. I was a columnist at the old *News American*. We chatted. It turned out Jim was well acquainted with my family. He and my father, Joseph B. Kelly, had been schoolmates at Loyola College where both names appeared in the "Greyhound," the college newspaper. Jim, of course, was staff photographer. Both remain Greyhound basketball fans.

Jim Gallagher says that as a photographer, he was blessed with luck. He rarely mentions skill and patience, and also the determination of the athlete he still is. He is a man with an energy level that never quits. When he sets out to make a shot, he will not stop until it conforms to the image he has precisely predetermined. If he has to go out seven days in a row for a certain shot, he does. If the train is four hours late, Jim is at his post. Others would have quit in frustration.

We worked together in 1983, when I was researching photos for a picture history of Maryland. Jim said he might have a few candidates. His "few" turned out to be a massive study of the Middle Atlantic region of rails in the 1950s. I was dazzled by the technical brilliance, the mood, and the drama of this collection. It was immediately obvious that work of this quality, most seen by only a few people, deserved to be enjoyed by an appreciative larger audience.

I realized that Jim's photography was unique in this part of Maryland, West Virginia, and southern Pennsylvania. First of all, Jim, ever the gentleman, was on very friendly terms with the management of both the Baltimore & Ohio and the Maryland & Pennsylvania railroads. These close relationships gave him a certain advantage — some of his best work was the result of his being tipped off, only a few hours in advance, that a certain locomotive would be moving through a precise location.

And to Jim, location was everything. Once, Jim received police permission to walk across the Route 40

Atmospheric rail shots don't always need a train. Here empty B & O branch line rails outside Harpers Ferry beg for attention.

Steam over the Susquehanna River.

Susquehanna River bridge. That occasion yielded a fabulous scene of a doubleheaded steam freight reflected in still waters (shown above). He would venture out in all kinds of weather, and even stand in dry — or wet — creek beds in order to get the best shot. His car trunk came fully equipped with thick winter leggings, coats, grass whips, saws, and other paraphernalia.

Jim also had occasional precious freedom to break loose on a work day. In the 1950s, he and his brother were overseeing the construction of Ednor Gardens, the distinctive Baltimore neighborhood just north of Memorial Stadium. His family developed the site beginning in the 1920s. On any given day, the B & O publicity department might telephone the construction shack to tell Jim that a particular locomotive would be leaving the Mount Clare shops.

He frequently jumped in his car and sped off to trackside with his Rolliflex and 4 x 5 "beat" Graphic. The film type he generally used was medium speed rated. He made fine grained, carefully developed negatives and processed his own prints in his home darkroom. He never used an exposure meter in his camera work, but relied upon his own experience and instincts for the proper exposure.

Jim began taking photographs while he was a Loyola High School student. Even while in college, a few of his photographs appeared in the old *Baltimore News-Post* and *The Sun.* Immediately after World War II, he made an extensive photographic study of Japan's Army and Navy air

forces. These 120 rare and historic prints, supported by Jim's writing, became the basis of a book, *Meatballs and Dead Birds,* now a much sought-after collector's item.

Beginning in the 1950s, Jim began to capture Baltimore photography awards. It was then he selected railroads as his special interest. A sampling of his work began to appear in *Trains* magazine, rail company publications, and *The Sunday Sun Magazine.* But these published works represented just a fragment of his railroad work.

Jim took a different approach to railroad photography. He knew that the 1950s were a major transition period for the Iron Horse. Some marginal lines were already on the brink of abandonment. Steam locomotives, although plentiful and technically powerful, grew more scarce by the month. It was his insight and gift to document that part of railroad change, when clouds of coal dust and deep steam whistles were still a part of a landscape now invaded by modern diesel locomotives.

Each of Jim's photographs tells a story and conveys mood, spirit, atmosphere, and character. The ground in his photographs rumbles. Some of the antique trains he photographed look as if they might rust and crumble before the end of their trip. Other photos impart a feeling of majesty and romance. And you don't have to be a train buff to appreciate them. Just step aboard and savor the results of Jim Gallagher's skill, luck, and persistence.

Jacques Kelly

WITH THANKS TO . . .

The late **Robert M. Van Sant** was the Baltimore & Ohio's director of public relations during the decade when Jim Gallagher shot the photographs in this book. Mr. Van Sant admired some of Jim's first published railroad work in *The Sunday Sun Magazine*. They subsequently met and a lasting friendship developed. More importantly, the B & O executive annually presented Jim with a press identification card which gave him unlimited visiting rights to B & O property. Otherwise, Jim could have been chased away by railroad police. Without the kindnesses Mr. Van Sant offered, this book would not have happened. He did not live to see this collection, but his welcoming spirit is very much in it.

B & O dispatchers, especially **Gene Scannell, Claire Fisher,** and **LaVere T. Neale,** contributed to this volume. **Danny Fisher,** the Viaduct tower operator at Cumberland, and **Garland Anderson,** Harpers Ferry station master, tipped Jim off to rail traffic movements.

The enthusiastic cooperation of **Arthur Gosnell,** store keeper at the Ma & Pa's Falls Road shops, was the key to photographing that line. Much appreciation also goes to **Harold Williams** who first published some of Jim's photos in *The Sunday Sun Magazine*.

Herbert H. Harwood, Jr., the dean of the B & O writers and historians, long encouraged the photographer and author in numerous ways. He generously read the manuscript and made excellent suggestions and corrections.

John P. Hankey of the B & O Railroad Museum also gave his time to study the photographs and point out many details of railroading history. His numerous suggestions are found in many of the captions.

Other individuals, **Daniel Sohn, Charles Hughes, Russell Cashdollar, Charles Tirschman,** and **James Genthner,** provided generous assistance.

And finally, the staff at Greenberg Publishing Company was warm and gracious. **Bruce Greenberg, Samuel Baum, Marcy Damon, Maureen Crum, Donna Price, Brian Falkner, Wendy Burgio,** and **Rick Gloth** worked to make this book a reality.

James P. Gallagher
Jacques Kelly

Cinders and coal smoke erupt from a mighty B & O articulated. The photographer saw a train in the distance, jumped out of his car, and shot the photograph from a West Virginia highway bridge.

Jim Gallagher at work alongside the B & O rails at Relay, Maryland.

Bathing the Budd car at Riverside yards, Baltimore.

A B & O engineer frowns at Jim Gallagher and seems to be warning, "Do not come any closer to my train!"

OF MEN AND STEEL

It takes much more than steel rails and moving iron to run a railroad. Railroads are huge industrial operations dependent upon dedicated and productive employees ready to work in shifts round-the-clock. Crews are up before dawn at roundhouses and yards, towers and shops. Steam locomotives especially demand constant maintenance coupled with hard physical labor. These photographs catch the rail worker with oil can and shovel — at the throttle and brake — all in a day's work.

On the morning of September 29, 1953, the smoke and steam shields the rising sun at the B & O's Brunswick yards in Frederick County, Maryland. A yard engine sits in the foreground, with the engineer and fireman, heads extended, silhouetted over the tender. Next in line is a mixed string of bay window and cupola cabooses. The switchman stands in the center, with the edge of a huge coal trestle at the far left.

Trackwork was still a manual occupation in the early 1950s when these two B & O workers set rail bolts. The process would soon be replaced by machines and continuous welded rail. The men who worked the track sections were known as "gandy dancers," a term that, by legend, derives from the workers' gait while walking on the ties. The rail in this photo was made by the Carnegie steel works in 1941 and is so marked.

Maryland & Pennsylvania (Ma & Pa) engineer John Whiteford has an added passenger in the cab of his diesel, August 18, 1952. Lady was the mascot of the Ma & Pa's Falls Road yards in Baltimore. Most shops kept a dog around.

The camera pans a fast-moving caboose while the eyes of the conductor stare into the camera's lens in July 1956 at Fairmont, West Virginia. The caboose has no electricity and appears uncomfortably hot.

The yellow lantern is lighted by kerosene. Someone has left the rearmost wheel's journal box open. The B & O built this caboose in January 1942 in Keyser, West Virginia from steel shapes and parts made at the railroad's shops in Cumberland, Maryland. The design, the bay window with the ribs and rounded roof, was unique to the line; these cabooses (and similarly designed boxcars) were called "wagontops" because they resembled old covered wagons. During the final years of the Great Depression of the 1930s, the railroad saved pieces of scrap metal, rolled them into shapes, and kept the work force busy by constructing railcars such as this specimen.

F. L. McGaha, whose white
shirt, tie, and hat mark him as
the B & O's assistant road
foreman of engines on the
Cumberland division, takes
the engineer's seat of an
eastbound diesel-powered
freight out of Cumberland.
His left hand tugs the air horn
at a Martinsburg, West
Virginia grade crossing on
June 30, 1954, while his right
rests on the air brake. For its
day, this was a modern
locomotive. If the cab slightly
resembles that of an
automobile, it should — it was
built by General Motors.

Here is
turn-of-the-century
railroading, where
man clearly conforms
to the demands of the
machine. Aboard the
Ma & Pa's work train
near Bel Air,
Maryland, engineer
Donald Hughes's
oil-stained leather
glove rests on the
throttle on September
12, 1952. He is aboard
the venerable No. 27, a
1906 Baldwin
Ten-Wheeler.

It is early morning at the Brunswick, Maryland yards as the day and night crews pass each other — one ready for work, the other for bed — on September 29, 1953. The light pole casts a long shadow over this atmospheric railroading scene, the crew change.

The ground these men walk on is covered with coal cinders. Some of the locomotives are pointed east, others west. The large steamer in the foreground hauls B & O passengers. The next three are freight engines.

The Pennsylvania Railroad operated in tight quarters in waterfront Fells Point, Baltimore. Operator C. G. Venzke, dressed in foul weather oil skins, climbs aboard this rubber-tired tractor used to shuttle freight cars around the vicinity of Jackson's Wharf at Bond Street in April 1952. The Pennsy built these specialized units in its Altoona, Pennsylvania shops specifically for the street trackage in Baltimore, and Jersey City, New Jersey. They were gasoline or diesel tractors, with a steel body and wood trim.

Part locomotive, part truck, the Pennsylvania's street tractors were familiar sights in the old wharf quarters of Fells Point, the 18th Century neighborhood along Baltimore's waterfront where the streets were paved with granite blocks and often crisscrossed with rails. They served export-import piers, waterside warehouses, and inland industries. These tractors did not have conventional steering wheels. Instead, they were guided by a wooden ship's helm. They had solid rubber tires on spiral-spoked wheels.

The engineer does some last-minute lubricating before starting a run out of the B & O's Riverside yards in South Baltimore, September 1953. He will soon be off to Washington. The oil can has a long spout for a reason — the engineer does not want to have to reach into the steam, dirt, and soot between the wheels.

Steam power dominates the B & O's ready tracks at Fairmont, West Virginia in October 1955. The men between the columns of locomotives oil up before the day's run. But lurking in the distance at the far left is a new General Motors diesel, waiting to take the thunder out of this display of steam might.

It is high noon at the B & O's Cumberland yards, September 30, 1953, but the smoke rising from the laboring Santa Fe-type locomotive 6161 throws the scene into temporary darkness. A heavy consist of freight cars proves too much for the big locomotive, so a diesel switcher is called in to help break the inertia. The little helper is hooked onto the front of the locomotive and supplies the required start. The diesel is then detached. Here it pulls away and eases onto a siding.

The steam and diesel locomotives each excelled at different tasks. The little diesel turned out its maximum power at low speeds; it could start a heavy train moving. The big steam locomotive, however, often had difficulties starting a train, but once moving, could pull the load at greater speed with less effort. The phrase that railroaders used was, "If you could start it, you could pull it."

The six men on the switcher crew are part of a drama that was enacted every day in railyards across the country.

Now under her own power, 6161 steams off to Pittsburgh. This is an impressive shot
of a steam locomotive working hard to get out of town and a classic illustration of
why these machines were the bane of anyone with a loaded laundry line.

A pair of shots of a freshly conditioned Pacific at the B & O's Riverside yards in Baltimore: this is where it all happens.

The steam, admitted to the cylinder, pushes the piston back and forth. All the power of the locomotive originates in this chamber. Here the steam is made into mechanical energy.

The driving rod is bolted to the wheel.

The power end of a B & O Time Saver fast freight as it moves eastbound at Savage, Maryland. The engineer, intent on his cigar, ignores the photographer in the pacing automobile in this June 1956 shot. This type of diesel was one of the most handsome locomotives associated with the railroad. It was streamlined and styled with a three-color paint scheme — black, blue, and gray with gold stripes and lettering, set off by stainless steel trim in the best 1950s fashion. Only passenger trains were faster than Time Saver freights.

The gateman at the busy Pennsylvania Railroad grade crossing at
Aberdeen, Maryland signals "Clear!" to the engineer of this rocketing New
York-Washington morning passenger train headed by a GG-1 in July 1954.
Note that the front pantograph is raised, an unusual practice. It is likely
that the rear pantograph was having trouble.

A wave from a man in a caboose captures the friendly side of railroading. The weatherbeaten motto reveals B & O's role, "Linking 13 Great States with the Nation." The man in the window is known as the hind-end man. He is doing what he is supposed to: looking out for any problems on the freight cars in front of him. The caboose is fifteen years old and appears twice its age. Its coat of devil's red paint and white lettering have seen plenty of hard work.

This show of locomotive power is impressive, although tranquil, in April 1951 at the B & O's Riverside roundhouse. The Pacifics, both built for the line's centenary in 1927, wear bells on their sides. Within the decade, all the locomotives of this class, except No. 5300, would be scrapped. No. 5300 survives at the B & O Railroad Museum at Mount Clare in Baltimore.

The Riverside engine terminal was so busy that the B & O had two identical roundhouses there. One was torn down in the 1950s, the other in the 1980s; they were replaced by the right of way for Interstate Highway 95.

Engineer and fireman board the B & O's Pacific-style locomotive in September 1953. She has just come out of the shop and sports a nicely polished running gear. Within a few minutes, the crew will pick up a string of passenger cars to become train No. 21, *The Washingtonian*, and serve waiting commuters along the Baltimore-to-Washington main line. The scene is at Riverside yards, Baltimore.

The yard hostler's job was to shift locomotives around the engine terminal.
Here one of the B & O's yard men relaxes for a moment inside the cab of a
B & O EM-1 class articulated, one of the largest locomotives the railroad
ever used. The view looks from the fireman's seat to the engineer. The
whistle cord hangs from the top in this July 1956 photo.

The Norfolk & Western was known for its modern steam fleet epitomized by roller bearings and super-power designs. The old oil can has been superseded by the Alemite gun, a heavy and bulky air-powered grease gun. It injects grease under pressure into the loco's roller bearings.

The comparative size of the worker gives an indication of the sheer bulk of this powerful freight locomotive, known for its tremendous pulling power, so ideally suited to the Alleghenies. The photo was taken at the N & W's Hagerstown, Maryland yards in July 1956.

The tiny Ma & Pa had its own railroad limousine, a 1937, eight-cylinder, eight-passenger Buick Roadmaster. Norman Spicer oils the car for railroad president Jack B. Nance, who later will be going out for an inspection tour. The rail adapter part of the vehicle was built in the Ma & Pa's Falls Road shop by Louis Miller and August Kramer in 1942. The car is equipped with air brakes, a whistle, and a two-way Bendix radio. It is now in the B & O Railroad Museum in Baltimore.

Water and coal ran a steam-powered railroad. On the early morning of September 29, 1953, a B & O tender takes a big gulp at Brunswick, Maryland. One of the first chores of the day is the trip to the penstock, where the water will be channeled into the tender. The hostler uses his right hand to guide a pulley to raise the water spout while his left hand operates the water valve.

OF MEN AND STEEL

Brother and sister, on the way to Sunday school, wave at the engineer of the B & O's *Ambassador* at St. Denis, Maryland. The train, moving in excess of a mile-a-minute, is accelerating out of a tight S-curve between Relay and St. Denis. It is just a few miles from Baltimore, where it will make a triumphant entry into Camden Station. The photo was taken in August 1952, when the train was pulled by diesel from Detroit to Washington and by steam to the end of its run in Baltimore. ◀━

This junior engineer has but one ambition. The engineer, whose leather gloves are coated with dirt, knows a different side of railway romance.

Even small lines followed railroad rules and traditions. The white flags on the front of Ma & Pa No. 27 signify the "extra" status of this work train. Its crew has performed railroad chores — maintenance and repair work. Here the crew pauses five miles outside Towson, at the Maryland School stop, to take on water from an old shingled water tower on September 12, 1952. After decades of hard service, the locomotive was retired three years later. Her pilot (railroaders did not like the term "cowcatcher") has been covered with sheet metal to help push away snow.

On a rainy November 2, 1955, Ma & Pa engineer Edwin E. Jones pours sand into No. 27 at the Bel Air station in Harford County, Maryland. The rails have become slippery and the lumbering loco needs traction. The old coal scuttle, which has sprung a sand leak, looks as ancient as No. 27. The other dome on the boiler collects steam. It seems to have a leaky safety valve. The 1906 Baldwin locomotive was soon retired after many years of climbing the hills of central Maryland and Pennsylvania.

It is early morning on a cool fall day at Riverside yards, Baltimore. The sun glances off the rich royal blue of fully coaled Pacific 5307 on the ready tracks. Bundled against the chill, B & O trainmen stand by in September 1953.

On the Northern Central route, a Pennsylvania passenger train, the *St. Louisan,* races through a Texas, Maryland grade crossing en route to Harrisburg, October 25, 1953. It was a measure of the competitive spirit between the B & O and the Pennsy that each would vie for the Midwest passenger trade. The B & O had the more direct route between Baltimore or Washington and St. Louis. The Pennsy's trains were at a disadvantage and had to travel from Baltimore to Pittsburgh, then down the "Panhandle" route to St. Louis.

A B & O diesel gets a bath at the Riverside yards' "laundry" in Baltimore, September 4, 1954.

A B & O steam switcher gets hosed down at Fairmont, West Virginia, July 1956. Heavy insulated gloves and the large handles on the steam hose give an idea of what this hot job involves.

It takes plenty of coal and water to make the steam to run a railroad. Every twenty or thirty miles a heavy freight train has to stop and take water. The fireman has to climb atop the tender and perform this task. Here water flows through the arm of a water tank. The photo's vantage point is atop a caboose. It is July 1956 and most American railroads already had converted from steam to diesel.

A Western Maryland Railway switchman needs plenty of elbow grease to throw the switch for his local freight near the Narrows at Cumberland, July 1957. The locomotive will soon pass under a low bridge. As a safety measure, there is a "telltale" overhead to warn of the limited clearance.

The fireman of B & O No. 5008 keeps a sharp eye out for trouble on his side of his passenger train bound for Baltimore at Relay, Maryland on a July evening in 1952. Crews work long and hard to maintain the track's ballast as neatly as this. The locomotive is fresh out of the shop, with clean graphite paint on the fireboxes and smokeboxes. The black locomotive is neatly lettered in yellow paint.

A B & O Santa Fe-type locomotive takes on coal at the immense concrete Cumberland tipple in July 1956. This powerful freight locomotive was known as a "Big Six" because all members of this class were numbered in the 6100 and 6200 series and, obviously, were big. It was a favored loco for the Cumberland-to-Pittsburgh run.

The B & O's behemoth articulated No. 7620 gobbles a diet of coal, water, and sand at Fairmont, West Virginia in July 1956. The locomotive is huge, but the concrete coaling tower is even larger. Bucket conveyors lift coal to its top. It is then fed by gravity through chutes until it drops into a loco's tender. To the right, the trussed outlines of the Monongahela River bridge are just visible.

A fireman applies the final touch to a B & O Mikado type at Fairmont, West Virginia in October 1955. The old steamers required maintenance, but rail crews knew their wants. Wiping the headlight was one of the hostler's jobs. The first Mikado-type locomotive (2-8-2) was built for service in Japan. Afterwards, American railroads named the 2-8-2-type for the Japanese character in the highly popular Gilbert and Sullivan operetta of the same name. The Mikado on stage was a power figure. So, too, were these locomotives. During World War II, the B & O tried in vain to rename them the "MacArthur," but the term did not stick.

The engineer can see this curve of coal hoppers from his cab at Flemington, West Virginia in July 1956. The freight conductor signals the engineer to edge up to a line of coal cars. The freight conductor "talks" to the engineer as if he has a two-way radio. Through hand motion and body language, he conveys the distance between the cars to be coupled.

This yardside Toscanini conducts the movement of a freight's passage at Port Covington, the Western Maryland Railway's deepwater freight terminal in Baltimore, October 1956. The switchman gives the go-ahead signal to pull ahead after two cars have been uncoupled.

A railroad man would describe this scene at the B & O's Riverside yards, Baltimore, as "hosing the grit," or dropping sand into the diesel (sand is used to improve traction on slippery rails). The hostler (the yard engineer who moves locomotives within a yard) looks on from the cab. There is still some soapy water on the windshield.

The Ma & Pa's No. 6 was a railroading legend in the 1940s and 1950s, when she was chronicled in magazine and newspaper stories. Built by the American Locomotive Company's Richmond works in 1901, she served on the line's Baltimore-to-York, Pennsylvania passenger runs for a good fifty years. This photo was taken at the Falls Road yards in Baltimore, April 11, 1952. She has been put on standby service, but is still ready to roll. Note her stack cap, the protective lid atop her smokestack used to prevent rainwater from entering her smokebox and corroding it. Four months later she was pulled to Patapsco Scrap, a subsidiary of Bethlehem Steel.

The Ma & Pa's section gang loads some new rail on a flatcar the hard way in March 1952. Some eleven good men had to have strong backs.

On May 16, 1954 at the Hagerstown, Maryland yards, a Norfolk & Western hostler lined up this impressive show of steam power for the cameraman. The N & W opted to use steam power later than most American railroads and seemed to delight in this decision. Most railroad workers, in those days, even had patience with photographers trying for a certain shot.

These proud Norfolk & Western streamlined passenger locomotives still had a few more years of service remaining when they were photographed in 1954.

A mechanic works on a massive Norfolk & Western locomotive at Hagerstown, Maryland in 1954.

POWER PLUS!

The physics of train movement is a complicated study of energy, motion, weight, and geography. Maryland's topography, from the flat Chesapeake basin to the rugged Allegheny Mountains, demands an impressive catalogue of pulling and pushing power. When one engine simply will not do the job, a dispatcher uses all the necessary might at hand to move his train.

Western Maryland Railway engineers called these diesel locomotive units "covered wagons." The four units with a mixed freight bound for Connellsville, Pennsylvania move into the Narrows along Wills Creek outside of Cumberland in July 1957.

Wills Creek joined the Potomac River at Cumberland but often, after major storms, overflowed its banks, flooding the town and ripping up railroad tracks. Several years previously, the Army Corps of Engineers had constructed this concrete channel to contain the water's temperamental moods.

The Western Maryland's Cumberland-Connellsville extension was planned by financier George Gould (Jay Gould's son) as a link in his grand plan for a transcontinental railroad.

A baggage cart defies the law of gravity at the B & O's Harpers Ferry, West Virginia station's platform in the 1950s. The cart is loaded with a mountain of merchandise shipped via Railway Express Agency (REA), the railroad industry's package shipment agency. The express company was owned jointly by the major American railroads and was a unified and predictable way of delivering goods, particularly heavy or oversized pieces that the post office would not handle. For example, the tires on the baggage cart are just the type of merchandise that Railway Express handled.

Many railroad stations, no matter how small, were Railway Express Agencies where goods could be picked up or dropped off for shipment. In larger cities, Railway Express trucks also delivered packages to their destinations. At this point, in the 1950s, REA was an important source of revenue for passenger trains that may have otherwise run at a deficit.

Three general purpose Western Maryland locomotives, attached to the rear of the mixed freight, give a lot more than just moral support. Once the freight is over the grade, this trio of helpers will uncouple and pull away. The Western Maryland was in close competition with the B & O for freight traffic. The Western Maryland assigned high horsepower per ton of freight hauled over this rugged country to insure dependable deliveries. The locomotives were dressed in speed lettering and stripes, designed to give the impression that the Western Maryland delivered the goods rapidly.

January 20, 1951 began as a clear calm morning, when a doubleheader of eastbound B & O Mikados (the mainstay of the B & O's Baltimore-Philadelphia freight operations) and mixed freight cars crosses the Susquehanna River, Garrett Island and approaches Aiken, Maryland. It is a rare moment when photographic conditions are this ideal. The still air helped to create a mirror reflection on the cold waters. A matched pair of steam locomotives is the miracle of railroad serendipity for which photographers have been known to pray.

Jim Gallagher is ready. He had already obtained permission from the police to shoot from the U.S. Route 40 Highway Bridge near Havre de Grace, Maryland. After he spots a plume of steam rising through the bare trees and hears the engines' pounding, he waits for the right moment to snap the photograph. Jim titled this photograph, "Doubleheader."

A giant Baldwin-built B & O articulated takes on sand at the Brunswick, Maryland yards. At the time, this class was among the most powerful of the steam locomotives in use east of the Rockies. Brunswick, a railroad town on the Potomac in Frederick County where the line built extensive yards and shops in 1890, was the farthest east the B & O took its big articulateds. No. 7612, seen here in October 1952, will soon be making the haul to Cumberland, Maryland and points west.

The engine is next to the yards' coal-fired powerhouse that generates electricity and steam in addition to supplying compressed air and hot water. In the background, there is a high water tower that feeds water to the penstock.

In a scene photographed three years later, October 12, 1955, the same No. 7612 takes to higher ground along West Virginia's Little Ten Mile Valley at Brown. This magnificent machine is moving a coal drag out of Fairmont and on to New Martinsville. Along the way it crosses this dirt road.

Fairmont was a gathering point for coal mined in two dozen different mines in the lower West Virginia panhandle. Not all the coal would be shipped eastward. Much of it was moved north to the B & O's Akron subdivision and ended up at Cleveland steel mills or the line's coal docks at Lorain and Fairport, Ohio.

A Ma & Pa doubleheader bound for York, Pennsylvania rolls out of the Falls Road yards in September 1952. In the distance is the great stone arch of the 29th Street Bridge. This old wooden boxcar pulled by the locomotives was for use only on the Ma & Pa. Because of the fragile car's age and wooden underframe construction, it could not be interchanged with other railroads.

A July sunset at the Hagerstown, Maryland yards lights this
Norfolk & Western Mallet. The man climbing into the locomotive is
photographer Jim Gallagher, who, helped by a ten-second self timer, wants
to try to make a more dynamic scene. Boarding a locomotive takes quick
wit, a cat's reflexes, and confidence.

Here is a mixed pair of B & O locomotives. The Pacific, the second engine, has just come out of the Mt. Clare Shops and is earning its keep by working west on the Old Main Line. The trip served to break in the locomotive on the steadily ascending grade through Carroll County. The heavy smoke shows these machines are working hard along the Patapsco River valley between Avalon and Ilchester. Until the turn of the century, this area had been heavily industrialized with tanneries, iron works, and mills. It was an important and early source of railroad revenue.

By 1952, heavier freights out of Baltimore had almost become the exclusive domain of the B & O's diesel locomotives. Doubleheaded steamers were quite scarce at that time. However, on September 5, 1952, Robert M. Van Sant, the B & O's public relation's director, called Jim Gallagher at about two in the afternoon and said, "There'll be a doubleheader freight moving west on the Old Main Line, leaving Baltimore around three." Jim left for Relay, Maryland. The train was late and he searched through Patapsco State Park for a place where the sun would shine the longest. His luck held. The train, an unusual Mikado and Pacific combination, broke through the shadows and cut through the only sunlit space in the area. His camera recorded black iron on the move.

North of Havre de Grace, Maryland, the GG-1s shoot under the catenary on the Susquehanna River Bridge in July 1956. The old bridge abutments in the foreground once supported another, earlier railroad bridge. This span was converted to automobile use until the U. S. Route 40 bridge was completed. Both the new highway and railroad bridges' decks were considerably elevated to avoid spring flooding.

The yard hostler moves the 7600-class articulated to a ready track at Fairmont, West Virginia in July 1956. The photo gives an indication of the block-long size of this monstrous piece of rolling machinery. It appears to be a simple photograph, but it took Gallagher several tries to get the portrait as he wanted it. None of the B & O's articulateds was saved from the scrapper.

The engine is covered with a fine layer of sand, especially around the wheels, which indicates that it has been laboring up a heavy grade. In 1956 the B & O was working these machines very hard. At this point, they were even being renumbered to make way for the arriving diesels. Although only twelve years old, she and her sisters would be retired in two years.

The engine is basically a large, complex steam power plant. In another form, it would be capable of lighting several thousand homes or running a dozen factories or powering a large ship through water. ◀▬

A pair of Pennsylvania Railroad GG-1s pound the Susquehanna River Bridge at Havre de Grace with a mixed freight speeding behind in July 1956. The GG-1s, introduced in the middle 1930s, were the long-lived workhorses of the electrified Pennsy main line between New York and Washington. Designed by industrial designer Raymond Lowey, they were virtually synonymous with speedy East Coast rail travel in this period. Painted Brunswick green with gold striping, the locomotives ran from New York to Washington for more than forty years, ending their lives under Amtrak and Conrail ownership.

The lead GG-1's open door gives a clue to the tremendous heat that its transformers radiate. On this hot summer day, the engineer is trying to catch a breeze. At the locomotive's center, the temperature is well over 155 degrees. Blowers attempt to cool the traction motors, but even when stopped at the end of a run, the GG-1 discharges heat. Engineers compared the locomotive to a thoroughbred horse that had to be cooled down after a race.

The diesel locomotive is clearly in the ascendancy in this October 11, 1955 shot at the B & O's Fairmont, West Virginia yards. Here three Western Maryland general purpose diesels move a train through the B & O's territory. The Western Maryland has trackage rights on the B & O's rails from Bowest Junction, Pennsylvania to Fairmont. The drag is a hundred odd empty hoppers destined for Chiefton. A B & O eight-wheel steam switcher, soon to be replaced by more efficient diesel-electric power, passes by the younger generation of locomotive power.

The first rays of the morning
sun on October 10, 1952
bounce off the station railing
at Harpers Ferry, West
Virginia as a pair of
articulateds approach the
Maryland Heights tunnel.
The locomotives are crossing
the B & O's bridge spanning
the Potomac River.

A helper locomotive, visible in the center with its trail of cottony white
smoke, assists a B & O steam locomotive to pull a long string of empty
hopper cars west from Grafton, West Virginia to Clarksburg. This scene,
taken the morning of October 12, 1955, is at Rosemont, West Virginia.

Six streamlined B & O diesel units pace along westward out of
Cumberland in July 1956. The shot was taken from Highway 220 near
Cedar Cliff along the Potomac River.

The skies are temporarily darkened at Simpson, West Virginia as a coal train, made up at Flemington, West Virginia, gets help from the local freight, moving east to Grafton, October 10, 1955. The thirty-five heavily loaded hoppers and the steep grade were too much for one B & O Mikado to handle. It stalled, or "hung up," literally unable to make the grade. After an hour's wait, a local freight showed up to get things going by coupling to the train's rear end and pushing. Even then it was no easy task.

There were no two-way radios for engineers — each 1,000 feet apart — to communicate with each other in this situation. All the movement was executed by whistles, hand signals, and operating practices. First, the local freight edged up to the stalled train's caboose and gently pushed forward, taking up the slack between the coal hoppers' couplers. The lead engineer blew his whistle when he was ready to start. The pusher then "heave hos," shoving cars forward to "bunch" the slack action. The lead engineer judged when to start by sound as the moving cars approached the engine. He then pulled away. The whole sequence is an action of finesse ruled by the laws of railroad physics.

On a hot day in July 1957, a B & O mixed freight moves eastbound through Luke, Maryland by the plant of the West Virginia Pulp and Paper Company. Brakeshoe smoke rolls out from some of the cars' wheels as the engineer has full control of the situation. The train is reaching the bottom of Seventeen Mile Grade having passed through rugged country. Note how the fumes from the paper mill have drifted up the hillside, killing the vegetation. Today, with environmental improvements, the hill is fully endowed with foliage and trees.

A real express train, a long string of baggage cars loaded with mail and package express shipments, is ten minutes out of Cumberland, working its way toward Keyser, West Virginia and the Alleghenies. The train, No. 29 on B & O schedules, left Cumberland at 8:40 a.m. September 28, 1953. In this era, the mail moved by rail. Mail-carrying contracts were an important source of railroad revenue.

A doubleheader steam drag of empty hoppers passes another freight at Relay, Maryland on an early August morning in 1952. You can almost smell the creosote on the ties. Jim is ready for this one. He hears the approaching steam locos as they whistle miles down the track at Halethorpe. But then a diesel-powered freight eases by, threatening to screen off his view. Fortunately, it clears just in time.

The two freights pass at Relay, Maryland.

A B & O Mikado 4417 puts on a real show of steam in October 1955. The locomotive heads a train of empty coal hoppers off a Flemington, West Virginia siding to the main line headed for Clarksburg, West Virginia.

The scene could have been planned by movie director Steven Spielberg. Vaporous steam billows from the turbo generator, safety valve, and piping, and a perfect cloud of smoke erupts from the locomotive's stack.

A pair of B & O steamers pulls a coal drag off the wye track at Flemington, West Virginia and heads for the yards at Grafton, October 12, 1955.

Jim Gallagher's final steam safari was April 12, 1958 to Roanoke, Virginia, where the Norfolk & Western was still moving some coal trains by steam power. Here a N & W coal drag rolls alongside the highway. This line was the last major railroad in the country to run steam in regular revenue service.

One can almost feel the ground trembling as the two powerful locomotives climb the upgrade along the western slope of the Blue Ridge.

Jim had only two days here, and coal movements were slow. One daylight run did take place on April 12, 1958. The shots were planned in advance. The woman wearing the white jacket is Jim's wife, Betty Gallagher.

It doesn't take a railfan to appreciate the excitement of seeing and hearing these majestic locomotives working the climb from Roanoke to Blue Ridge Summit. Here helper No. 2152 gives a lift to a diesel-powered freight.

Two hardworking N & W Mallets break the tranquility of the countryside as they storm by a small farm house.

A glorious full side view as No. 2142 thunders by.

The fireman of No. 1210 takes a look at the camera as the two engines battle the upgrade at about 15 miles per hour. This three-quarter view of the doubleheader catches another train approaching at the distant grade crossing. The spent steam collects along the right of way like dense ground fog.

The camera is elevated on an overpass as the coal train is about to pass. It may be April, but there are still snow deposits along the rails.

At full throttle, these giant N & Ws plow through the Virginia country on this raw April day.

Smoke shooting heavenward, the N & Ws roll through a signal interlocking on this overcast afternoon. The car following the first locomotive's tender is a "canteen," the railroad's term for an auxiliary water tender.

OVER . . .

The railroad surveyor's challenge becomes the photographer's blessing. Maryland is crisscrossed by countless creeks, streams, and rivers. The resourceful railroads found ways to bridge the obstacles. The natural geography of the Potomac and Patapsco valleys provided the routes west. Along the way, the stone and steel spans imparted a visual drama that afforded a unique vista for the skilled photographer.

On a warm September morning, the sun cuts through the mist over Wills Creek at Cumberland Narrows in western Maryland.

A diesel locomotive eases over the Carrollton Viaduct in southwest Baltimore, July 24, 1954. The oldest railroad bridge in the country spans Gwynns Falls west of the B & O's Mount Clare shops and was part of the road's original right of way out of the city. The span was completed in 1829 and has one main arch over the water, with a secondary arch for an old wagon road. The main arch is 52 feet above the water and the entire span is 312 feet long. Designed by James Lloyd, it was built under the supervision of Caspar Wever, one of the B & O's early construction managers. The shot shows scaffolding used for routine painting and maintenance.

OVER . . .

Bound for Camden Station, the B & O's *Ambassador* charges across the Thomas
Viaduct at Relay, Maryland on a frosty November day in 1952. The
Ambassador was a deluxe train out of Detroit headed by a dark blue
President-class locomotive, consisting of two Pullmans, two standard coaches,
and a baggage-lounge car. The Thomas Viaduct which crosses the Patapsco
River valley was named for the B & O's first president, Philip E. Thomas.

About one hundred years separate the masonry bridge from the three
streamlined passenger diesels heading the *Capitol Limited* into Baltimore. The
lush growth in the Patapsco River valley indicates the lateness of the summer,
August 25, 1952. The locomotive's smooth features contrast nicely with the
rough Patapsco River quarry stonework on the curved Roman-arch span.

Spring rains in May 1952 have caused tree limbs and debris to collect at the northern base of the Thomas Viaduct at Relay, Maryland. The Pacific has been rebuilt, with its center driver replaced. The first car behind the tender is a World War II troop sleeper converted into an express car.

A Baltimore-bound commuter train passes over the Thomas Viaduct, September 30, 1952. The freshly overhauled locomotive is not long out of the shops and the first passenger car has not yet been painted the line's gray and blue color scheme. Also note that it has sealed windows thanks to air conditioning. The morning sun accents the trim cast-iron railing. The Patapsco River has silted forty more years since this photo was taken and as a consequence, the river bed is higher today.

Designed by Benjamin Henry Latrobe II (son of the United States Capitol's architect), the Thomas Viaduct was constructed like an ancient Roman bridge. In two-thousand years, the technology had not changed. Even though challenged by the Patapsco countless times, the bridge is as strong today as the day it was built. Its massive piers sit directly on bedrock and its wide, graceful arches that shoot high off the valley floor allow trains to cross, even at critical flood stages.

The span is the first multi-span masonry railroad bridge in the country and the first to be built on a curving alignment.

A Ma & Pa four-car train rolls northward through the Cromwell Valley of Baltimore County toward Bel Air, Maryland on a rainy February day. The whole line was but 80 miles long, with 114 bridges accounting for 11,155 feet.

Ma & Pa's locomotive No. 41 pulls a boxcar and four-wheel caboose over the timbers of the 450-foot long Gross Trestle near Sharon, Maryland in 1955. The photo shows the span's distinct curve, an unusual structure on an unusual railroad.

The Ma & Pa's No. 27 is silhouetted on the Long Green Trestle as the freight winds its way through the countryside toward Baltimore on November 2, 1955.

The engineer of the Ma & Pa's No. 27 locomotive salutes the camera as he eases this train over the trestle near Loch Raven, Maryland, November 2, 1955.

The Pennsy's 1930s electric workhorse, the GG-1, rockets efficiently over the Gunpowder River bridge. Pennsylvania trains have little charm; they compensate with cold efficiency. The scene is vaguely reminiscent of a Florida Keys publicity shot. The picture, taken in September of 1954, illustrates the might of the Pennsy at that time. The 1935 electrification project has been superimposed upon a bridge built years before. The coaches span the World War I era to the lightweight streamline era.

Leaning into the curve, this B & O Santa Fe-powered
mixed freight charges eastbound through the Narrows at
Cumberland, Maryland, October 9, 1955. Fast and
powerful brutes like these were standard motive power
on the east end of the Pittsburgh Division.

A coal drag charges through a downpour at Monogah, West Virginia, October 12, 1955.
Photographer Gallagher's 1939 Packard contrasts in scale to a Q4-class Mikado. Note the
differing construction standards on the highway and railroad bridges.

Arthur Gosnell, the Ma & Pa's storekeeper, let photographer Jim Gallagher
know that locomotive No. 43, a Baldwin 2-8-0 built in 1925, was going to
have an Interstate Commerce Commission inspection in Baltimore and
would be traveling the length of the line. Jim set up his camera and waited
hours in a dry stream bed, South Stirrup Run, watching the Gross Trestle
near Sharon, Maryland in Harford County. It began to shower and he was
about to quit and go home. Then he heard a steam whistle in the distance.
The changing weather brought perfect atmospheric conditions. The
trainman steps outside on the caboose's rear platform and the scene is set
for this superlative railroad photograph, November 10, 1955.

A glorious scene from the Hilltop House at Harpers Ferry, West Virginia catches a trio of B & O Budd cars crossing the Potomac in September 1960. Since that time, the scene has changed very little and the RDCs continue to operate into the 1990s. Various efforts to remove an old sign that reads "Mennen's Borated Talcum Toilet Powder" (barely visible on the side of the cliff left of the tunnel) have failed.

A B & O President-class Pacific moves out of Harpers Ferry for a run into Washington, D. C., October 18, 1952. The Railway Post Office car behind the tender gives an idea of how important mail traffic was to railroad revenues. Mail patrons could walk up to the car and hand a letter to one of the men who worked this important communications link. James Gallagher's camera catches the secondary plume of steam trailing over the engineer's cab. The condensation is the product of a turbine that generates electric power for the locomotive's headlamp, cab, and marker lights.

B & O train No. 34 heads for the tunnel at Maryland Heights, October 18, 1952, over a bridge built high enough to be protected from floods. The engine is about to pass over the C & O Canal which follows the Potomac River north all the way to Cumberland. The sun glances off the gray, blue, and gold passenger car sides in a moody atmospheric shot.

This fisherman's eye view of a diesel-powered, eastbound *Capitol Limited* was shot beneath the B & O's Susquehanna River bridge, September 2, 1954.

On a February 1955 day, the old Patterson Viaduct abutments on the B & O's Old Main Line supported only telephone and telegraph wires that were part of the line's communication system. This 1831 stone Patapsco River bridge, part of the original route from Baltimore to Frederick, was destroyed in an 1868 flood and replaced by an iron truss, which in turn was supplanted by a tunnel and new bridge not far from this site. The Patterson Viaduct was named after William Patterson, one of the line's original directors.

Passengers regularly looked forward to the ride over the Susquehanna River, literally one of the high points of East Coast runs of both the B & O and Pennsy. Here the B & O's *National Limited* does its high line act on a crisp September morning in 1954. Garrett Island foliage completes the picture.

A westbound Western Maryland Railway freight destined for Connellsville, Pennsylvania rolls across a steel bridge over U. S. Route 40 just west of Cumberland in July 1956. By contrast, an older masonry span crosses Wills Creek. Lettering on the side reveals the name Cumberland and Pennsylvania Railroad, a short line that served mines in the Cumberland area.

Just as the September 1954 sun pops over the eastern horizon, a morning New York-bound Pennsylvania Railroad flyer makes quick work of the Susquehanna River bridge at Havre de Grace. The lighting at this hour reveals the stone ice-breaking masonry work on the north side of the bridge abutments, a touch of engineering that passengers could not see from their seats.

The B & O's *National Limited* will be in Mount Royal Station in Baltimore shortly, but there is still time for main line scenery as the train crosses the Susquehanna River. The afternoon sun highlights bridge construction details as well as the engine's paint job. The bridge was completed in 1909 and opened in early January 1910.

An energetic steamer backs a string of empty coal hoppers on a local move across the Potomac River at Harpers Ferry, October 5, 1952. The forested hillside in the background is Maryland Heights. The tracks in the foreground are part of a local rail siding. Those in the middle lead to the branch up the Shenandoah Valley to Winchester and Strasburg Junction, Virginia.

Jim Gallagher recalls that it was a pretty October 1956 day when he planned to get some Western Maryland Railway power shots at the Spring Garden drawbridge over the Middle Branch of the Patapsco River in South Baltimore. He found only two shots worth taking. A mixed freight had been made up in the line's sprawling Port Covington deepwater yard and taken off for Hagerstown. He shot it twice, once as it initially negotiated the timber trestle part of the double-track span and later as its engines arrived at its steel draw. The bridge, across an old shipping channel, connected Port Covington with Westport, site of the Baltimore Gas and Electric Company's power plant. Ⅲ➡

. . . UNDER

No railroad trip seems complete without an excursion through a dark tunnel. These subterranean channels, the construction of which tested the prowess of surveyors and engineers, trim travel time considerably. Some underground cuts go through hills and mountains, others pass under highly populated and busy parts of Baltimore. A good photographer can make use of the lights and shadows in these rail caverns for interesting effect.

A fast-moving B & O mixed freight, led by diesel 949, erupts from the Ilchester, Maryland tunnel onto a steel span across the Patapsco River, heading west on the Old Main Line in August 1952. The masonry inscription "19-ILCHESTER-03" is visible above the tunnel opening. The tunnel replaced an earlier alignment on the B & O Old Main Line.

A hallmark Pennsy keystone emblem on a modified P5a electric locomotive breaks into the sunlight at the Union Tunnel at Greenmount Avenue, Baltimore. Both the B & O and Pennsy worked their ways underground through the most populous parts of Baltimore. The construction of the Union Tunnel, Baltimore & Potomac tunnels, and the B & O's Howard Street tunnel took many years and consumed a major amount of capital.

Ma & Pa locomotive No. 30 labors up the steep grade through Baltimore's Stony Run Valley in Wyman Park, on the west side of the Johns Hopkins University campus. The northbound train will soon pass through the University Parkway bridge's soot-blackened concrete arch, May 1952.

An eastbound Western Maryland fast freight exits a tunnel leaving south Cumberland en route to Baltimore on September 30, 1953.

The B & O's 1896 Mount Royal Station in Baltimore is located in a bowl at Mount Royal Avenue and Cathedral Street. Trains arrived or departed the station via the Howard Street tunnel, built between 1890-1895, to give the line a more direct route through Baltimore. Here, on June 3, 1954, the Diplomat has just pulled out of the station and noses into one of the tunnel's arched granite-clad apertures.

A steam-driven freight train heads west into the tunnel at Point of Rocks, Maryland, October 1, 1953. It has just passed the station at Point of Rocks and pulled off the Old Main Line. The scene is in Frederick County, southeast of Brunswick. The name derives from the place where the Catoctin Mountains meet the Potomac River.

Jim Gallagher has his lens ready within the tunnel at Maryland Heights, just across the Potomac River from Harpers Ferry, West Virginia. A B & O coal drag clatters across the bridge, October 5, 1952.

Granite walls, heavy concrete foundations, and huge steel bridge girders accentuate the rugged power of the mighty B & O articulated No. 7625 on October 5, 1952. It has just passed Harpers Ferry, West Virginia, rolled across the Potomac River, and is about to make the ground tremble as it enters the Maryland Heights tunnel. It pulls a mixed freight. This big freight engine traveled no farther east than Brunswick, Maryland. There it was put on the turntable and pointed west.

NIGHT-LIGHTED

When the sun goes down and the day has ended, the trains still run. Railroading goes on round-the-clock for the people who work the rails. Night crews perform rail housekeeping tasks so that all is ready for the next day's commuter, long-distance passenger, and freight hauls. The camera here goes after the dark and dramatic side of railroading seldom seen by most railfans.

Ma & Pa's old No. 27 rolls over the timber trestle on the grounds of the Shepherd and Enoch Pratt Hospital, near Towson, Maryland, November 2, 1955. The camera's flash bathes the pistons of the old locomotive in a ghostly light on one of her last runs. She was sold for scrap in January 1956.

Jim Gallagher lugged a 4 x 5 graphic camera and tripod as he climbed up
the B & O's Riverside yards coal tower, Baltimore, for this night shot of
B & O Budd cars at the fueling station in July 1956. Come morning, the
cars will begin their Baltimore to Washington commute at Camden Station.
Note that the lead unit is a combine.

After a day's work, on the night of November 9, 1953, the Ma & Pa's No. 27
coals up at the Falls Road timber dock. The site today is part of the
Baltimore Streetcar Museum's operation. Almost forty years later, some of
the old timberwork remains in place.

B & O Alco diesel No. 828 stands alongside Pacific 5066 at
Riverside's ready tracks, October 24, 1953, in Baltimore.

The B & O was a stickler for cleanliness. Here passenger locomotive No. 86 gets freshened up with an evening shower at Riverside yards, Baltimore, August 1954.

Two power competitors square off on Riverside yards' ready tracks
November 3, 1953. The General Motors Electro Motive Division's (EMD)
unit is on the left; on the right, an American Locomotive Company's (Alco)
product. In the center, a mechanic checks the blackboard locomotive roster
posted on a crew shack.

A Norfolk & Western Mallet stands ready to depart Hagerstown, Maryland during a heavy rain in October 1954. Hagerstown was the northern terminus of the N & W's Shenandoah branch. The Mallet will pick up its consist at Vardo and head south through the Virginia coal country.

B & O Alco and General Motors diesels rest up by moonlight at B & O's Riverside yards, July 1955.

Two B & O Pacifics frame a just-risen sun at Riverside yards, Baltimore, September 24, 1953.

It is still night, but the first morning rays have broken through to light a B & O doubleheader at Harpers Ferry on the Strasburg branch in October 1952.

A silhouette in steam appears at Relay Bend, Maryland, July 1951.

Passenger, general purpose, and road freight locomotives stand at attention at Riverside yards, Baltimore, August 1954.

Roy O. Stuart, the Ma & Pa's night foreman, drops the hot coals from the firebox of No. 27 at the Falls Road yards, Baltimore, November 2, 1955.

This night scene at the Ma & Pa's Falls Road yards contrasts the overlapping technology of forty years. Diesel No. 81, built in 1946, stands alongside the pride of 1906, No. 27, November 9, 1953.

At the end of the day, after the fire has been put out by dropping the hot coals into the ashpit, there is still enough steam left for No. 27 to power herself into the roundhouse.

Charles Armbuster checks out the massive Norfolk & Western's K2 No. 130 in the train shed at Hagerstown, November 26, 1954.

It sometimes takes sand to start on slippery rails. Here a Budd car takes on its supply at Riverside yards, Baltimore, August 1954.

A B & O commuter Budd car's polished stainless steel sides get a soapy bath in August 1954 at Riverside yards, Baltimore.

Bathed in incandescent light, a General Motors diesel
drinks its fill at Riverside yards, November 3, 1953.

RUNAROUND FOR LOCOMOTIVES

Railroaders recognize the names Brunswick, Fairmont, and Riverside among others. These are the railyards with their huge brick roundhouses and turntables where the steam giants literally got the "runaround." Railroad companies did not encourage casual visitors around these massive pieces of moving machinery. Here is a side of railroading that passengers seldom glimpsed.

B & O Pacific 5306 stands on the turntable at Riverside yards, Baltimore, October 24, 1953. This locomotive made the B & O's last scheduled passenger steam haul from Baltimore, going to Cumberland and back as Trains 21 and 22, on November 2, 1953. The fire was put out that evening at 8:30 at Riverside.

The Ma & Pa's Falls Road yards in Baltimore had a hand-operated turntable, known in railroad jargon as an "Armstrong," because of the physical labor it took to move a locomotive. One of the line's diesels gets turned in October 1955.

The hostler and table operators pose B & O Pacific 5059 on the Brunswick, Maryland turntable, September 30, 1954.

When the table operator and hostler are working for you, more than one picture is called for. Another view of the 5059.

The Ma & Pa's Falls Road yards were often an operating museum of antique steam. Here two railroad employees use a dose of elbow grease to move No. 27. After the railroad discontinued operations here, the site was taken over by the City of Baltimore's Department of Public Works. The locomotive stalls were then used by the city's fleet of winter salt trucks and snowplows.

Also at the Ma & Pa's yards in November 1955, Jim shot from the perspective of No. 30's cab, a 1913 Baldwin 0-6-0 switcher being turned about.

Norman Spicer changes the direction of No. 29 at the Ma & Pa's
Baltimore yards in March 1952. Built by Baldwin in 1913, the engine
served until 1956.

Plumes of steam and smoke rise from the B & O's Brunswick, Maryland roundhouse in October 1952.

For her return trip to Baltimore, the Ma & Pa's motor car pauses on the York, Pennsylvania turntable in July 1954. Note the arm of the man at one of the center windows.

A fireman's eye view as a mighty B & O articulated is turned on the table at Fairmont, West Virginia. The hostler is about to ease his engine into the roundhouse on this July 1956 day.

A spanking clean B & O 5500-series "Mountain-type" rolls slowly onto the big Brunswick, Maryland turntable in October of 1952.

B & O articulated 7612 is given the runaround to place it in the roundhouse. This gigantic powerhouse of a locomotive, with tender, leaves only a few inches to spare on the table's rails, October 10, 1952.

TRANSITION

In looking at the history of railroading from the days of the Tom Thumb to the present, the themes of progress, change, and transition arise. Never was American railroading so affected by technology as it was in the relatively short period after World War II until the middle 1950s. This was a decade of revolutionary change. The coal-fired iron horse was replaced by the highly efficient diesel locomotive. Lens and film record the passage of the steam boiler and its replacement by diesel motor.

The replica of the B & O's Tom Thumb and an open passenger car pass Halethorpe in the middle 1950s, about the time that diesel power had come to dominate the picture. Steam reigned for nearly one hundred twenty-five years in the mid-Atlantic region. This shot was taken at a television show rehearsal.

Jack Nance, president of the Maryland & Pennsylvania Railroad, boards his train in Baltimore for its last passenger run on August 31, 1954. The last run of the Ma & Pa was an occasion — there were many additional passengers. The local press even covered the event.

The Ma & Pa rented an extra Pennsy coach for its last passenger run. Instead of a normal consist of a motor car and baggage trailer, diesel 81 pulled the line's pair of motors, the borrowed Pennsy day coach, and a baggage car, to accommodate those who wanted to make the final trip from Falls Road to York, Pennsylvania. Within four years the Maryland portion of the line would be abandoned.

Within two years, steam power would vanish from the B & O's
Grafton, West Virginia yards. But on October 10, 1955, steam and
diesel shared the same stage.

An early, wet snow in November 1952 at Relay, Maryland dresses the landscape for this pair of photos. The Washington-bound *Ambassador* steams along the B & O's eastbound main line in a scene that would not be repeated very often.

A few minutes later, the *Capitol Limited* appears, pulled by shiny new diesels. The village of Relay, in Baltimore County, takes its name from an earlier form of B & O power. In the early days of the line, B & O coaches received a fresh relay of horses here.

Railroad companies needed replacements for steam locomotives in the 1950s. A popular vehicle for this was the Budd Company's Rail Diesel Car (RDC). Self-propelled, the car could run alone or in multiples. The B & O bought a fleet of these Budds and called them "Speedliners." Railfans often referred to them as "the Budd cars." Here train No. 161, a Baltimore-to-Washington commuter, is about to pass under the automobile overpass at St. Denis in September 1953.

The B & O's mighty steam era was a memory by late 1953 in the Baltimore area. The Riverside roundhouse held nothing but diesel power at the time of this September 4, 1954 photo. The RDCs would stay in service into the 1990s.

Older steam locomotives surround a Baldwin "shark nose" diesel at the B & O's Grafton, West Virginia yards on a cool rainy July morning in 1956. The Baldwin road diesel was not too successful in service and was rather scarce on the B & O. Their tenure was far shorter than the steam they replaced.

The old and new order pass each other at Relay, Maryland on a frosty November 1951 morning. A speeding diesel takes the outside of the curve while a steam-powered freight moves in the opposite direction. The sun has not had time to melt the frozen dew on the crossties. With a bit of extra good luck, Jim Gallagher's camera catches the passing of a steam-drawn local freight, Washington-bound, opposite a diesel-pulled passenger train, Baltimore-bound. His film records a pair of trains, moving at speed, a rare treat for the camera-carrying railfan.

Railroads felt the acute pinch of auto, bus, and air competition in the 1950s. The Pennsylvania Railroad experimented with this General Motors Aerotrain, a streamlined affair with matching passenger coaches. The Aerotrain was no faster than normal trains and certainly less comfortable, but its main virtue was operating economy. Here it is photographed near Halethorpe, January 5, 1956, on a special run from Washington, as it passes under the B & O tracks.

A steam locomotive shuttles some coal hoppers on the wye near
Flemington, West Virginia as a B & O diesel rolls by in October 1955.

Only a few days before this November 6, 1953 shot was taken at Riverside
yards, Baltimore, the B & O made its final scheduled steam run out of
Baltimore. The early snow dusting only accentuates the stilled, cold boilers.
By the next month, no steam locomotives would remain at once-smoky
Riverside. Some would be scrapped immediately; others would be sent west.
Note the replacement diesel's nose at the left as it edges into the scene.

Helping to usher in the modern railroad era, the Pennsy's sleek Aerotrain wings through Halethorpe, Maryland toward Baltimore at 90 miles per hour on January 5, 1956. It was later put on the Philadelphia-Pittsburgh run.

PASSING STATIONS

Running concurrently with the demise of the steam locomotive was the passing of the passenger station. Once the pride of a village, town, or city, these depots fell victim to sharp competition from automobiles and airlines. Railroads once congratulated themselves on the design and character of their stations. But by the 1950s, when these shots were taken, many wayside and city stops were at the end of their service. Scenes of people congregating to board trains would soon become an uncommon sight as travel by air and on interstate highways surged.

The ground quakes as a pair of B & O Pacifics race by the blacktopped platform and its bishop's crook light standard at St. Denis, Maryland.

All are aboard and the conductor signals the engineer for the *Marylander* to leave Mount Royal Station, Baltimore, October 1956. The passenger car's paint job and cleanliness are faultless, in keeping with B & O standards.

A B & O Mikado heads a mixed freight off the Old Main Line westward at the Point of Rocks station (built between 1871-1875) in Frederick County near the Potomac River, October 18, 1952. Rail historian Herbert H. Harwood has referred to this classic depot as an "exhilarating conglomeration of shapes and polychrome masonry culminating in a high cupola and spire topped with an iron finial, it is the prototype High Victorian Gothic railroad station."

A fireman looking back over the B & O's *Metropolitan Special* would view this scene at the Point of Rocks station, June 30, 1954. The wood towers, which controlled train operations, were a familiar sight along the line.

In a few minutes, the Pennsy's *Gotham Limited* will be at Baltimore. These diesel units, with their "cat's whiskers" gold striping, have just passed the Lake way station (near Lake Roland) in Baltimore County. The train left Harrisburg, Pennsylvania earlier in the afternoon and moved south along the old Northern Central right of way, October 25, 1953.

The Pennsy's Riderwood Station in Baltimore County was a commuter station, but intercity trains on the Northern Central line regularly passed its doors. It was a typical PRR suburban depot in an area that was served by commuter rail service until 1959. Today the line is part of Baltimore's light rail right of way.

A Ma & Pa switcher No. 30 eases under the great stone arch of the North Avenue bridge in Baltimore. The line's Baltimore station is the simple wood building at the left, reached by several flights of steps. The line once possessed a Romanesque-style depot, but it was torn down in 1937 to allow construction of the Howard Street bridge.

The Ma & Pa's aged but obviously still reliable No. 27 was pressed into service November 10, 1953, when the line's pair of gas-electric motor cars were not running well. Here she works through the Towson, Maryland station and is ready to take the steel span over York Road on her way to Bel Air, Maryland and final destination of York, Pennsylvania.

The Ma & Pa's Bel Air, Maryland station was a sleepy county stop on this warm September afternoon in 1951.

On a drizzly November 2, 1955, the Ma & Pa's No. 27 picks up three freight cars on a haul from York, Pennsylvania to Baltimore. The old switch and station signal blend right in with this scene of railroad antiquity. The locomotive was retired shortly after the picture was made. In less than three years, the rails would be ripped up and nearly all the Maryland portion of the line abandoned.

Jim Gallagher tries for an old-time railroad publicity shot. Bright August sunlight floods a Washington-to-Baltimore commuter run at St. Denis, Maryland.

Passengers step down from an efficient Budd RDC at St. Denis, Maryland in May 1954. B & O train No. 155 will call at Washington and Brunswick, Maryland.

Iva E. Crowthers, the Ma & Pa's station agent at Glenarm, Maryland, waits for a freight to edge past her Baltimore County station. Locomotive No. 27 rolls by with its haul on a chilly November 2, 1955.

Conductor Paul D. Hanes picks up his orders from Mrs. Crowthers on the fly from his four-wheel caboose. The brakeman gets his portrait taken as he sits in his vantage place.

A fast-moving westbound freight pounds by the B & O's Ilchester, Maryland station, August 22, 1952. The station agent, in white shirt and straw hat, checks the rolling stock, his eye out for journal hot boxes.

"On time!" Two B & O trains meet at Harpers Ferry, West Virginia on the Potomac River. The westbound *Metropolitan Special*, on its Washington-to-St. Louis run, rolls into the station as steam-powered No. 34 prepares to exit, October 18, 1952, 9:20 a.m.

The lure of the steam whistle is hard to resist. A B & O local steam freight hauls through St. Denis in August 1952.

A General Motors Electro Motive Division diesel handles an eastbound freight past the Ellicott City, Maryland station in the winter months of 1955. The old fieldstone house in the center looks about as ancient as the B & O's 1830 station, once an important early destination of the line.

The majestic *Washingtonian*, bound for the District of Columbia and points west, stops at Mount Royal Station on a cold January 1952 day. She is about to enter the Howard Street tunnel portal and roll under Baltimore's busy department store district. Note the trolley wires and poles on the street above. The fireman looks back for the conductor's signal to move on and plunge into the shadows. This particular locomotive, the *President Washington*, No. 5300, is preserved at the B & O Railroad Museum in Baltimore.

A Washington-to-New York diesel passenger train, the *Capitol Limited*, howls out from Baltimore's Howard Street tunnel to reach Mount Royal Station's platform and covered train shed in January 1952. After decades of battling the Pennsy, whose Baltimore station was only a few blocks away, the B & O decided to give up passenger runs to Philadelphia and New York in 1958. The station itself ceased to operate in 1961 and was sold to the Maryland Institute College of Art. The Howard Street tunnel still resounds with a full schedule of freight traffic.

An American Locomotive Company's diesel works west along the B & O's Old Main Line past the Ellicott City station's platform in January 1951.

The head end of a lengthy B & O fast freight eases into the Cumberland, Maryland station in July 1956. Unit 956 is the first of six diesels pulling this consist west via Grafton, West Virginia.

The cast-iron, vine-patterned balconies of the Cumberland's Queen City Hotel frame a parade of diesels pulling the westbound *Cleveland Night Express*, July 1956. The old railroad hotel was demolished in the 1970s after a lengthy debate and cries of protest that it should have been preserved.

A Baltimore-bound doubleheader passes by on the Old Main Line.

SCENE THROUGH A WINDOW

It was in the summer of 1950 when Jim Gallagher got the notion that railroads would be his primary photo subject matter. A friend told him about Relay, Maryland, a small community outside Baltimore adjacent to the B & O's Thomas Viaduct. One July Saturday he found the place. The first thing that struck his eye were the remains of the old Viaduct Hotel, an 1872-1873 building constructed in the "Y" formed by the Washington Branch and the Old Main Line. This once-grand hostelry, with walls of Patapsco River valley granite, was being razed. Jim started looking around and spotted an old window as a prime location for some shots. The window had an eccentric shape. It was set into a picturesque Victorian slate-trimmed tower at one end of the hotel.

Rain washed out his first effort. He knew he had to get back there fast before the place disappeared. His next date was the first Saturday in August 1950. It was an extremely hot day. Jim stood with his camera on a tripod for several hours. He realized there would never be another chance here — even the roof had already been taken off. The last shot of this small series of photos made the cover of *The Sun's* rotogravure magazine, November 9, 1952.

Mixed freight and hoppers rumble by.

A passenger train approaches.

Then another freight rolls toward the Viaduct Hotel window.

Both passenger and freight trains diverge. This ideal photo came early in Jim's rail photo sessions. The result only whetted his appetite for more. The odds of capturing two trains, nicely framed, and spotted in just the right places are high. The shot is a testament to a rail photographer's patience.

A Ride on the Ma & Pa

About 7:40 p.m. on August 31, 1954, a diesel switcher pulled a short string of passenger coaches along Falls Road near the North Avenue Bridge in Baltimore. About one hundred people stood on the span and watched the proceedings. Another group watched from the tracks. The train was the last regularly scheduled passenger run of the Maryland & Pennsylvania Railroad, the Ma & Pa. The Ma & Pa was a beloved, 19-mile-an-hour relic of railroading. It rambled — on a course that might have been designed by a lost cow — through city and county, field, stream, hill, and wood. Its locomotives were ancient; its passenger cars unashamedly unstreamlined. It was a 1900s antique that the gods of smoke and iron preserved well into the atomic era. It was the railroad you loved to love.

The final day of passenger service was cool and clear, the air purified by Hurricane Carol, which had moved on to New England. The evening sun caught the returning day coaches and outlined them brilliantly for their final working hours. In a few months, a baggage car and day coach would go off to transportation museums.

The United States Post Office had helped keep the line's passenger service alive (one train a day from Baltimore to York, Pennsylvania, and return) via an annual $25,000 mail-hauling contract. But the government terminated that business as of September 1, 1954. Passenger revenues the previous year amounted to a paltry $4,136. The line applied to the Interstate Commerce Commission for permission to abandon passenger service. The ICC ruled in its favor. The last trip was posted for August 31, 1954.

That morning's mail-passenger train, which had been made up in Baltimore and rambled the Ma & Pa's tracks, passed by its respectable stone engine house, broke through the foliage into Wyman Park, hugged the Stony Run valley through Roland Park, and rolled by stations and flagstops called Evergreen, Notre Dame, Woodbrook, Sheppard-Pratt, Towson, Oakleigh,

Loch Raven, Notch Cliff, Glenarm, Long Green, Hyde, Baldwin, Laurel Book, Fallston, Vale, Bel Air, Bynum, Forest Hill, Rocks, Street, Pylesville, Whiteford, and on to York. The route was both circuitous, twisting up and downhill, and notoriously slow. The line wound around so much that a passenger could always see where the train was going long before the locomotive got there. It took the panting and puffing freight engines four hours to make York from Baltimore. What they lacked in speed they compensated for in sooty character.

In the 1950s the Ma & Pa roster was filled with old steam veterans of countless mail and milk runs. These locomotives and coaches would have been ancient in the 1930s. By 1954 the line's Falls Road roundhouse was an unintentional, eloquent working railroad museum. The neighboring yards were oil-stained, covered with ash and coal dust, and were absolutely oblivious to the 20th Century. There were wobbly sidings of shabby wooden boxcars and seldom-used day coaches. When the City of Baltimore built the Howard Street Bridge in the late 1930s, the main Baltimore station sat in the span's path and was demolished. The railroad never got around to building a new terminal. Passengers had to descend a rickety flight of wooden steps to a plain, unadorned platform. The Baltimore homestead of the Ma & Pa was tucked away on a narrow ledge of the Jones Falls valley, just to one side of the North Avenue bridge's eastern buttress.

Jim Gallagher spent July 2, 1954 with his camera shooting a typical day in the life of the Ma & Pa's train No. 1, the mail run out of Baltimore. He picked up a schedule, a card about 2½ by 7½ inches in size, and began his record of eleven hours in the life of a soon-to-vanish passenger railroad. Freight operations continued for a few more years, but the Maryland portion of the line disappeared completely by June 1958. Work crews ripped up the rails and the old stone roundhouse became a shed for city street-salting trucks.

TIME TABLE

In Effect SUNDAY, APRIL 25, 1954

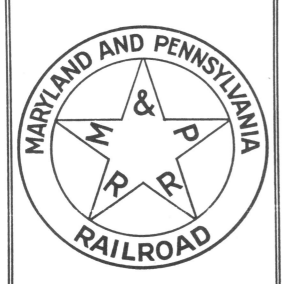

MARYLAND AND PENNSYLVANIA RAILROAD

M & P R R

SUBJECT TO CHANGE WITHOUT NOTICE—
This Time Table shows the time at which trains may be expected to arrive at and depart from the station named, but their arrival or departure at the time stated is not guaranteed nor does the Company hold itself responsible for any delay or any consequences arising therefrom.

A. M. BASTRESS - - Traffic Mgr.
J. B. NANCE - Pres. and Genl. Mgr.

TIME TABLE
EASTERN STANDARD TIME

Effective

April 25, 1954

Read Down 1	Miles	Daily, except Sunday STATIONS	Read Up 2
Leave A. M.			Arrive P. M.
6.40	0.0	..........Baltimore..........	5.20
f 6.47	2.6	..........Evergreen	f 5.06
f 6.50	3.2	..........Notre Dame......	f 5.04
f 6.54	4.3	..........Homeland	f 5.01
f 6.56	5.1	..........Woodbrook	f 4.59
f 6.58	6.2	..........Sheppard..........	f 4.57
s 7.05	7.0	..........Towson..........	s 4.55
f 7.08	7.7	...Towson Heights..	f 4.43
f 7.12	9.6	..........Oakleigh......	f 4.38
f 7.16	11.2	..........Loch Raven......	f 4.34
f 7.17	11.9	..Maryland School.	f 4.32
f 7.18	12.6	..........Summerfield......	f 4.31
f 7.20	13.4	..........Notch Cliff......	f 4.29
s 7.26	14.5	..........Glenarm......	s 4.27
f 7.30	15.8	..........Long Green......	f 4.23
s 7.33	16.8	..........Hyde..........	s 4.20
s 7.37	18.4	..........Baldwin......	s 4.17
f 7.43	21.2	......Laurel Brook.....	f 4.09
s 7.49	22.3	..........Fallston......	s 4.06
f 7.53	24.2	..........Vale......	f 4.00
s 8.05	26.5	..........Bel Air......	s 3.55
f 8.10	28.5	..........Bynum......	f 3.45
s 8.16	30.3	..........Forest Hill......	s 3.41
f 8.21	32.3	..........Sharon......	f 3.32
f 8.24	33.7	..........Fern Cliff......	f 3.29
s 8.30	35.3	..........Rocks......	s 3.25
f 8.38	37.3	..........Minefield......	f 3.19
s 8.42	38.6	..........Street......	s 3.16
s 8.46	40.3	..........Pylesville......	s 3.11
s 8.52	42.4	..........Whiteford......	s 3.06
s 8.55	43.3	..........Cardiff......	s 3.02
s 9.00	43.8	..........Delta......	s 3.00
f 9.05	45.9	......Bryansville......	f 2.50
f 9.08	46.9	..........Castle Fin......	f 2.47
f 9.14	49.4	..........Southside......	f 2.41
s 9.18	50.6	..........Woodbine......	s 2.37
s 9.21	51.6	..........Bridgeton......	s 2.34
f 9.26	53.5	..........Bruce......	f 2.29
s 9.33	56.5	..Muddy Creek Forks	s 2.22
s 9.36	57.1	..........High Rock......	s 2.19
s 9.42	59.4	..........Laurel......	s 2.13
f 9.46	60.9	..........Fenmore......	f 2.09
s 9.48	61.3	......Brogueville.....	s 2.08
s 9.54	63.6	..........Felton......	s 2.02
f 9.57	64.8	..........Brownton......	f 1.57
f 10.02	66.8	..........Springvale......	f 1.53
s 10.12	68.3	..........Red Lion......	s 1.50
s 10.19	70.5	..........Dallastown......	s 1.38
s 10.27	70.3	..........Yoe......	s 1.28
f 10.28	70.8	..........Relay......	f 1.26
f 10.31	72.1	..........Ore Valley......	f 1.23
f 10.32	72.8	..........Ben Roy......	f 1.21
f 10.33	73.3	..........Enterprise......	f 1.20
f 10.35	74.4	..........Paper Mill......	f 1.18
f 10.37	75.3	..........Plank Road......	f 1.16
10.50 A. M. Arrive	77.2	..........York	1.10 P. M. Leave

f—Flag Station.   s—Stop.

Trains will stop at stations marked "f" on signal or notice to Conductor. Presentation of tickets to these stations will be considered sufficient notice.

Writer Robert Breen once described the Ma & Pa's mail cars as rambling gypsy wagons. Here, in the early morning hours, the York-bound mail sacks are transferred via motor truck from the Pennsylvania Railroad, which had an interchange track with the Ma & Pa along Falls Road in Baltimore.

This covered frame staircase was part of the Ma & Pa's "new" Baltimore terminal which replaced its proper granite station, a victim of the Howard Street Bridge project. The staircase, which descended from the southern side of the North Avenue Bridge, was inconspicuously marked for the passerby.

The Baltimore waiting room at North Avenue was a simple frame structure at the shadow-filled base of the Jones Falls valley under the stone viaduct. The line was still hauling the occasional milk can in 1954, as it had been doing for the previous seventy years. The Baltimore terminal was unpretentious, but it was located squarely in the heart of a major league railroading neighborhood. A short distance away was Pennsylvania Station, in the 1500 block of Charles Street, and the Baltimore & Ohio's Mount Royal Station, at Mount Royal Avenue and Cathedral Street. Both of these lines had heavy traffic through the immediate precincts of the Ma & Pa yards. The B & O's trains passed directly overhead, while the Pennsy's Northern Central division paralleled the Ma & Pa's property for several hundred yards just across the Jones Falls stream bed.

Conductor Paul D. Hanes takes his twin daughters, Phyllis and Paula, and their older sister, Frances, on one of the last Ma & Pa passenger trips. (Passenger business had picked up slightly after publication of the closing date.) The girls are all wearing wrist watches. The Ma & Pa, like other American railroads, observed standard time, even in summer months. The Ma & Pa had its quirks. Most railroads operated morning commuter trains to bring suburban residents into the city. This line went the other way, taking domestic workers to jobs in the suburbs.

Motor car 62, with engineer E. E. Jones at the controls, is about to leave Baltimore, 6:40 a.m. EST. The distillate electric car was built in 1928 and saw heavy service on the line. The railroad found these units more economical to run than the coal-fired steam locomotives.

Conductor Hanes showed many little kindnesses toward his passengers. He was careful in calling out stations and flagstops along the way, but one passenger has missed her call at Long Green. He nevertheless stops the train at a nearby country road to discharge her. The conductor later explained the road stop was more convenient for the rider anyway. The Ma & Pa was a railroad of many little courtesies.

Just eighteen miles and nearly an hour out of Baltimore, the train passed through the gently rolling meadows at Baldwin. The cows along the way had been a source of income to the railroad, with milk shipping once a major factor in the line's revenues.

Some passengers get help boarding at Laurel Brook, a Harford County flagstop on the Little Gunpowder River. They will probably go only as far as Bel Air, a little more than five miles away and for a few quarters in fare. Laurel Brook is a favorite local summer waterhole. When the Ma & Pa train returns this way in the late afternoon, the engineer gives a sharp tug of the whistle. That blast signals the end of swimming day.

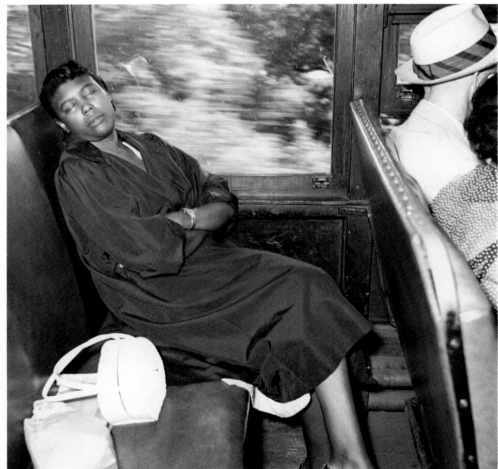

While some passengers are out to sightsee, others ride the line to get to work. The train's leisurely speed and its motor's monotonous throb lull this rider into the Land of Nod.

At Bel Air, Train No. 1 makes a full station stop. J. Hunt Stafford and
his grandson Tucker O'Connor stand on the wooden platform before
boarding No. 62 for a day's outing while the opportunity is still
around for a passenger ride.

Engineer E. E. Jones handles the controls of No. 62 as it leaves Bel Air. "It was a rare treat to be up front in one of the Ma & Pa's gasoline-electric motor cars. This photograph captured a moment of this experience," Jim Gallagher said.

The Ma & Pa provided its riders with many delightful views. The line had sixty-nine bridges, the largest of which was the Gross Trestle near Sharon in Harford County. This photo, taken from the engineer's cab, shows the timber span and its sharp curve over a dry creek bed.

The station at Rocks was characteristic of the scenery as the line meandered through the countryside in search of milk cans, varied merchandise, and the occasional passenger. Before trucks finally came to dominate the freight business, the Ma & Pa worked hard to cultivate the local shippers. The railroad may have been a bucolic rambler, but there was a good economic reason for its existence.

Inside motor car No. 62, passengers appear unworried about the impending loss of passenger service to take place in a little more than a month. Interestingly, a Japanese firm issued a brass miniature scale model of this car in 1975.

When the train makes its scheduled stop at Cardiff, it straddles the Maryland-Pennsylvania state line, in the heart of the slate deposits — a major source of freight traffic. The conductor glances to see if anyone is coming to catch his train. By 1954, more often than not, there were no passengers.

After passing Delta, Pennsylvania, the Ma & Pa's roadbed followed
Muddy Creek for the next fifteen miles toward York. The line crossed over
bridges which spanned shallow, stony waters several times in a single mile.

Even those who thought railroading merely a means of getting between two points would agree that the Ma & Pa rivaled the best on scenic display. The fertile York County countryside was calendar-perfect, dotted with barns and farmhouses.

At Dallastown, Pennsylvania our train pulls off the main line for a fast brake inspection. The conductor and engineer will board the car and head on to York after making sure that there is no equipment problem.

A lone fare boarding at Yoe, Pennsylvania receives proper attention from
Conductor Hanes. Single fares traveling only a few miles made the Ma & Pa
something of a country bus.

For the return trip back to Baltimore, motor car No. 62 has reversed direction.
It furnishes compressed air to supply power to an ingenious turntable device at
York. Used here is an 8-inch brake cylinder from an old boxcar. A three-way
valve functions by drawing air from the train line. The cleats are made of oak.

During a two-hour layover at York, mail bags are transferred from a motor truck to the Ma & Pa's baggage car. Trucks will soon take over this source of revenue from the railroad.

Come 1:10 p.m., passengers at the York station climb aboard the Ma & Pa train No. 2 for the trip southward to Baltimore. The railroad used a black and gold color scheme for destination signs.

While a Ma & Pa trip lacked modern amenities, it possessed an affectionate informality.

It was hard to predict how and what kind of freight would arrive. The line's baggage-mail car No. 35, seen here, is a relic of 1906 and is preserved at the Baltimore & Ohio Railroad Museum in Baltimore. Conductor Hanes's posture seems to be saying, "What's coming in now?"

There was plenty of cold water in the tank, but a paper cup cost a penny. Here Paula Hanes yanks the pull on the dispenser.

It is 3:55 p.m. in Bel Air and there are no takers for the trip into Baltimore. A few Railway Express parcels are loaded, however.

How did Jim Gallagher get this shot? The Ma & Pa was a line that accommodated its riders in a most gracious manner. There was always something cordial, personal, and wonderful about this old road. Jim asked about the possibility of his own photo runby on a trestle near Towson. He thought it might make an excellent shot.

The engineer and conductor agreed. He was let off at the south end of the trestle and the motor car backed up. Train No. 2 rolled once again over that section of rail. Jim snapped the picture and climbed aboard. He wanted the last shot of the series to be a memorable one. This is the result.

CHASING LOCOMOTIVE 4442

Jim Gallagher recalls a trip he made in West Virginia during the waning days of steam:

"There is a phase of train photography in which I dabbled a couple of times. My basic approach was to plan, set up, shoot, and return often if that instant did not produce results to my satisfaction. I also did some train chasing.

"There are some very serious fans of this type who will literally hop in and out of their automobiles and photograph a given train repeatedly, at different spots, for trips measuring in dozens or hundreds of miles. Shouldn't orchestrating this approach and executing the chase properly be called a challenge or even a sporting endeavor?

"On one train-shooting safari on October 13, 1955, with my friend Gordon Timbs, I did a series of shots of B & O Mikado 4442 for about fifteen miles — not a long-distance chase in some photographers' view, but still fun. We found out that a steam-drawn mixed freight from Martinsville, West Virginia would be on the move through an area called the Little Ten Mile Valley. Highway Route 20 parallels the tracks. We had a good hour of planning time to check over the area for the best spots to shoot. The cloudy weather turned into heavy rains and started to look bad for photography. As my usual good luck would have it, all this heavy rain and the chilly temperatures actually helped the result."

"Just off the highway at Smithfield, West Virginia, B & O No. 4442 struggles along a steep grade through the valley in the fog and rain."

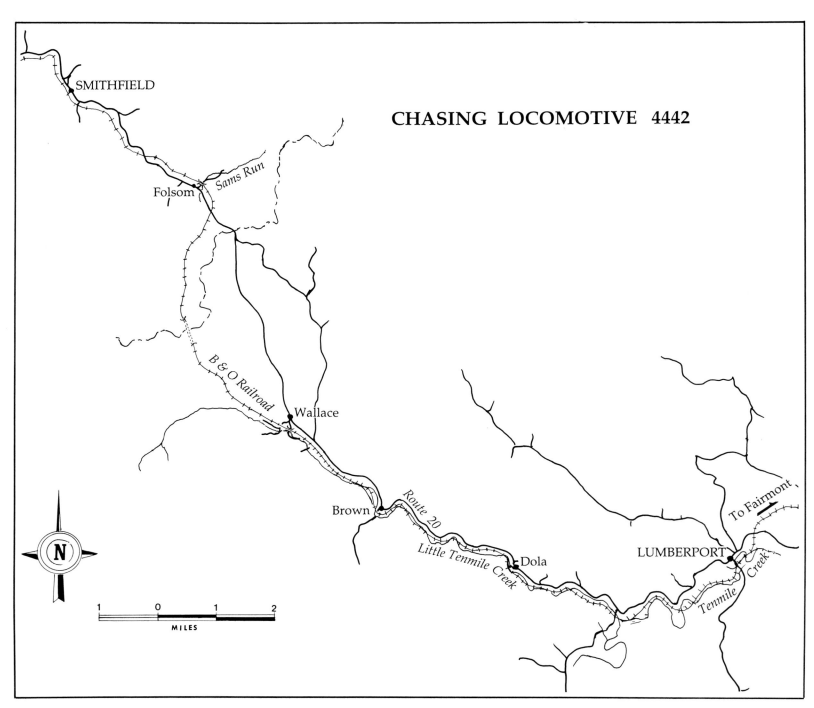

This map shows the route followed by Jim Gallagher as he pursued the B & O's
4442 from Smithfield to Dola, enroute to Fairmont, West Virginia.

"The worsening weather conditions help intensify the show of smoke and steam."

"Minutes later we catch No. 4442, still working hard upgrade as its storms across the highway west of Folsom, West Virginia. The locomotive's wheels churn the heavy rains on the tracks and roadbed, making a steam-like mist that adds to the action."

"The rains continue as No. 4442 crosses Route 20 again, this time at Hartzel, West Virginia, on an overpass bridge east of Folsom."

"For a while we follow No. 4442 along the highway at Wallace, as she is almost silhouetted against the rainy sky. Our engine continues to work against a rather steep grade as she passes a small lumberyard and sawmill."

"The smoke from No. 4442 drifts overhead as the train moves along the countryside and leaves the proximity of Highway 20 near Brown, West Virginia."

"We raced ahead of No. 4442 and her consist to catch them at a dirt road grade crossing at Dola. Here I combined passing Americana, the covered bridge, and the steam train. Gordon Timbs watches as the locomotive moves toward my camera. The train is on a downgrade now and speeds along toward its next stop at Fairmont. We gave up the chase here."

TRACKSIDE SCENES

It is hard for a rail photographer to resist the call of whistle, horn or bell. A fine fall day in the Potomac River valley or a cold and wet February afternoon in Bel Air, Maryland — the weather is always perfect if the train is running. In the 1950s, the wait was never too long for the patient trackside photographer.

The Ma & Pa's Old No. 27 winds along its crooked path in Maryland's Harford County countryside, November 11, 1955. She was then the oldest active steam locomotive in service out of Baltimore. At this time, all of the Ma & Pa's boxcars were so antiquated they could not be exchanged with other lines.

The engineer of the B & O's No. 34 obliges the photographer as he signals,
"Watch my smoke," at Harpers Ferry, West Virginia on October 1, 1953.

A B & O Pacific pulls away from a coal tipple at Riverside yards, Baltimore, with
the afternoon sun directly behind the structure in April 1951. Complicated railside
architecture such as this became a casualty of dieselization.

A back-lighted locomotive and tender, headed for the U.S. Route 1 overpass, clears the St. Denis, Maryland station in October 1952.

A morning Washington-to-Baltimore commuter train blasts a picturesque geyser of steam and smoke into the cool October skies as it hustles away from the St. Denis, Maryland station in 1952. The old day coaches had seen years of service. While the B & O ultimately eliminated long-distance passenger service, daily commuter demand on the Baltimore-Washington corridor remains high to this day.

A B & O Santa Fe heads a Pittsburgh-bound freight through the
Cumberland Narrows, September 30, 1953. The camera is at the edge of
Lover's Leap on Will's Mountain. Western Maryland Railway trackage,
Will's Creek, and highway U.S. Route 40 parallel the B & O here.

The roaring power of this three-unit, diesel-drawn freight resounds as it pulls
a long consist around Relay Bend in Maryland. The noise of steel against steel
breaks the silence as the train moves through the rail's sharp curve.

A speed marker a little east of St. Denis, Maryland frames the B & O's
Cleveland Night Express charging toward Baltimore, March 22, 1957.

Coming toward the camera is a man, just off the commuter train, who smiles at the events of this carefully orchestrated gag shot made in July 1952 at St. Denis, Maryland. Jim Gallagher is the tardy traveler.

A spotless B & O Pacific heading the westbound *Ambassador* rolls along an equally tidy roadbed at Relay, Maryland during the twilight of steam, the summer of 1951.

Daybreak at the Harpers Ferry, West Virginia station, October 1952. Just beyond the trestle, the Shenandoah and Potomac Rivers merge. To the left is the tunnel under Maryland Heights.

B & O locomotive 4443 moves a coal drag through Lumberport, West Virginia on October 12, 1955. The train which was made up at Martinsville heads toward Fairmont. The scene resembles a section of scale model railroading.

A B & O Baltimore-to-Washington commuter blasts a column of steam and smoke at St. Denis, Maryland, October 1952. Locomotive 5090 puts on the show.

The morning of September 29, 1953 reveals that steam is still the ruler of the B & O's Brunswick, Maryland yards.

Seldom has the mood and atmosphere of old-time railroading been captured as
vividly as in this morning scene at Brunswick, Maryland, September 1953.

The B & O's *Diplomat* rounds a bend on its Washington-to-Baltimore course in February 1955. The photograph was taken from a highway bridge at St. Denis in the Relay section of Baltimore County.

The eastbound *Capitol Limited* rolls downgrade along the Potomac River
valley about a mile west of Harpers Ferry, West Virginia in September 1957.

The junction of Relay with the small stone monument in the center shows
the site of the old Viaduct Hotel. The rails to the right are the B & O's
Old Main Line to Ellicott City, Maryland and points west. The train on
the left heads from Baltimore to Washington in October 1952.

Fleecy clouds seem to hang over the catenary at a spot about a mile northeast of Aberdeen, Maryland as the Pennsy's *Keystone* slips along at about 85 miles per hour. The run linked New York and Washington. The *Keystone* was an experimental lightweight train.

In a bucolic West Virginia setting, a helper on a coal drag works through Simpson headed for Grafton on October 12, 1953.

A panned shot of a Pennsy GG-1 accentuates the speed of this famous
locomotive type as it passes through Baltimore on its way to Washington.

A Ma & Pa light freight rambles through Harford County's rolling hills in November 1955.

The drowsy commuters on this cool, September morning in 1952 probably had no idea Jim Gallagher was shooting their trip to Washington.

The coal mines near Flemington, West Virginia provided the traffic for these
B & O locomotives in October 1955. The day's early light enhances the mood.

A B & O diesel moves upgrade along the narrows at Cumberland, along Wills Creek, pulling a string of empty hoppers to Connellsville, Pennsylvania in October 1955.

The fireman of B & O's No. 4555 takes a broom and housecleans his steam engine at the Fairmont, West Virginia yards on October 12, 1955.

Time was running out for scenes such as this. The Brunswick yard was dominated by steam locomotives on September 29, 1953.

The late Robert M. Van Sant, who served as the B & O's Public
Relations Director, arranged for Jim Gallagher to ride in the cab on
this and several other occasions. Here eastbound train No. 34 passes
near Harpers Ferry, West Virginia on June 30, 1954. Jim's shot was
taken from the cab of the B & O's No. 11, the *Metropolitan Special*.

Here a short Ma & Pa train crosses Charles Street at Woodbrook, Baltimore County, September 12, 1952. This photo was taken from the rather unusual vantage point of the caboose, the Ma & Pa's No. 2002.

Internationally recognized photographer, A. Aubrey Bodine, admired
Jim's rail photography. He especially liked this shot of the B & O's
Washingtonian on the move through Relay, Maryland in May 1954.
Jim used a slow shutter speed on the fast-moving train.